**W9-BQU-520**

# Farm Animals
# Sheep

### Heather C. Hudak

Weigl Publishers Inc.

Published by Weigl Publishers Inc.
350 5th Avenue, Suite 3304, PMB 6G
New York, NY 10118-0069
Website: www.weigl.com

Library of Congress Cataloging-in-Publication Data

Hudak, Heather C., 1975-
  Sheep / Heather C. Hudak.
      p. cm. --  (Farm animals)
  Includes index.
  ISBN 1-59036-428-7 (hard cover : alk. paper) -- ISBN 1-59036-435-X (soft cover :
alk. paper)
  1.  Sheep--Juvenile literature.  I. Title.
  SF375.2.H83 2007
  636.3--dc22

                                        2005034670
Printed in the United States of America
1 2 3 4 5 6 7 8 9 0  10 09 08 07 06

**Editor** Frances Purslow
**Design and Layout** Terry Paulhus

**Cover:** One pound (0.5 kilogram) of sheep's wool can be made into 10 miles
(16 kilometers) of yarn.

**Photograph Credits:** David Sanger photography / Alamy: page 7 TR

All of the Internet URLs given in the book were valid at the time of publication. However,
due to the dynamic nature of the Internet, some addresses may have changed, or sites
may have ceased to exist since publication. While the author and publisher regret any
inconvenience this may cause readers, no responsibility for any such changes can be
accepted by either the author or the publisher.

Every reasonable effort has been made to trace ownership and to obtain permission to
reprint copyright material. The publishers would be pleased to have any errors or omissions
brought to their attention so that they may be corrected in subsequent printings.

# Contents

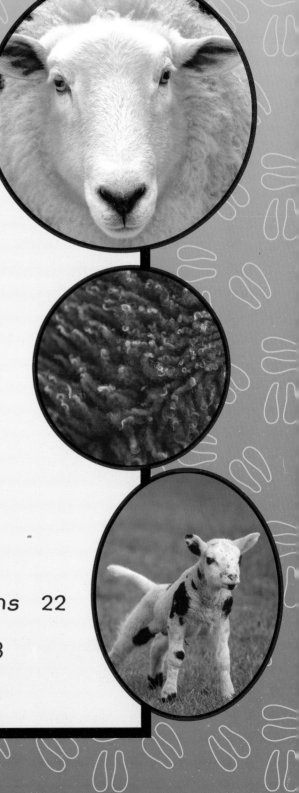

# Meet the Sheep

Sheep are small farm animals. They have long **muzzles**, pointy ears, and short tails. The thick hair that covers their bodies is called fleece. Some sheep have horns. Sheep come in many colors, shapes, and sizes.

Sheep are **mammals**. Mother sheep feed their young with milk from their bodies. Sheep are social animals, too. They stay in large groups called flocks. Sheep can remember one another even after years of separation.

Sheep are **grazing** animals. They spend most of their time eating grass in large, open fields called pastures. Grazing sheep also eat weeds. This helps prevent pastures from being overrun by unwanted weeds.

Like cows, sheep produce milk. Sheep's milk is used to make **gourmet** cheese.

Sheep are social animals. It is rare to see a sheep alone.

# All about Sheep

Adult male sheep are called rams. They sometimes butt heads with one another. Rams show off to win the attention of female sheep.

Sheep are easily frightened. Living in flocks helps protect them from **predators**. Sheep predators include mountain lions, coyotes, and wolves. In flocks, sheep will follow a leader. Where one sheep goes, the others follow.

There are many types of sheep. Different types of sheep produce different kinds of wool. Each type of sheep also has many **breeds**.

There are about 800 breeds of sheep.

# Types of Sheep

## Fat Tailed

- Used for milk or meat
- Coarse, long wool
- Found in dry parts of Africa, the Middle East, and Asia
- Breeds include Awassi, Bakhtiari, and Karakul

## Fine Wool

- Used for wool
- Thin or fine wool
- Found in Australia, New Zealand, South America, and the United States
- Breeds include Merino

## Hair Type

- Used for meat
- Has hair, not wool
- Found in Africa, the Caribbean, and some parts of Canada
- Breeds include Black-bellied Barbados and Blackhead Persian

## Mutton Type

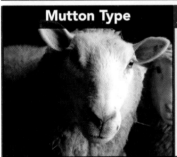

- Used for meat
- Medium to long wool
- Found in the United States and other parts of the world
- Breeds include Suffolk, Hampshire, and Dorset

## Short Tailed

- Used for breeding
- Medium wool and short tail
- Found in Scandinavia
- Breeds include Finn, Shetland, and Romanov

## Long Wool

- Used for wool
- Thick long wool
- Found in Britain, New Zealand, and the Falkland Islands
- Breeds include Lincoln, Coopworth, and Romney

# Sheep History

Sheep have been around for a long time. They were first tamed in Asia more than 10,000 years ago. Early sheep had long, rough, brown hair. Over time, this hair became soft fleece that was white, black, brown, red, or gray. People learned to spin fleece into wool 5,500 years ago.

In the 1500s and 1600s, English **settlers** brought sheep to North America. All sheep today come from the **urial** of southern Asia or the **mouflon** of southern Europe.

Sheep have many uses. They are used for clothing, meat, and leather. Lanolin is an oil found in sheep's wool. It is used to make candles and makeup, such as lipstick.

## Fascinating Facts

Queen Isabella of Spain used money from the wool **trade** to pay for Christopher Columbus's trips to the Americas.

People that raise and care for sheep are called shepherds.

# Sheep Shelter

On farms, sheep can live indoors or outdoors. Some farmers keep their sheep outdoors all year. Outdoor sheep need a shelter to protect them from poor weather. This shelter should protect sheep from strong winds, rain, and snow in winter and very hot weather in summer.

There are different types of shelter for sheep. They can live in barns or metal buildings. Shelters called hoop houses have an arch. The arch is covered with a heavy fabric.

Sheep shelters should not be heated. They should have many openings to let in air. Sheep that live indoors need bedding. Straw, hay, wood shavings, wood chips, and sawdust are good bedding for sheep.

Sheep are gentle animals. They are easy to care for. Like cats and dogs, sheep like to be around people.

Bedding provides warmth and comfort to indoor sheep.

# Sheep Features

Sheep are **hardy** animals. Their bodies are well **adapted** to the environment. Wool keeps sheep warm in cold weather. Muscular legs help them run quickly and easily. Other parts of sheep bodies also have special features and uses.

**TAIL**
Sheep are born with a long tail. The tail is shortened to help prevent illness and infection.

**WOOL**
Most sheep have white wool. White woolly sheep may have brown or black faces and legs. Wool can also be black, gray, silver, brown, or red.

## SENSES

Sheep have excellent hearing and a good sense of smell. They also have good eyesight. These senses help sheep know if predators are nearby.

## HORNS

Some sheep have horns. Some horns are hollow. They usually grow throughout a sheep's life. Some horns do not grow to their full size. These are called scurs. Sheep that do not have any horns are called polled.

## FEET

Sheep have cloven-hoofed feet. Their hooves are divided into two toes. Sheep also have skinny ankles.

# What Do Sheep Eat?

Sheep are herbivores. This means they eat plants, such as grass, clover, weeds, leaves, and twigs. Sheep eat plants that grow close to the ground. This is where young, tender plants grow.

Sheep graze for about seven hours each day. They do most of their eating in the morning, late afternoon, and at sunset. When there are no fresh grasses to eat, sheep eat hay. They also eat grains, such as corn, wheat, oats, and barley.

**Climate** affects where sheep eat. In dry places, sheep need to travel greater distances to find food and water. Farmers in dry places may need 10 acres (4 hectares) of land for 10 sheep. In a wetter place, a farmer may need only 1 acre (0.4 ha) of land to feed 10 sheep.

Sheep are ruminants. This means that a sheep's stomach has four parts.

During winter, sheep are brought indoors, where they feed on hay and grain.

# Sheep Life Cycle

Mother sheep are called ewes. Baby sheep are called lambs.

Ewes carry their babies in their belly for about five months. Giving birth to lambs is called lambing. Ewes can have one to three lambs at a time.

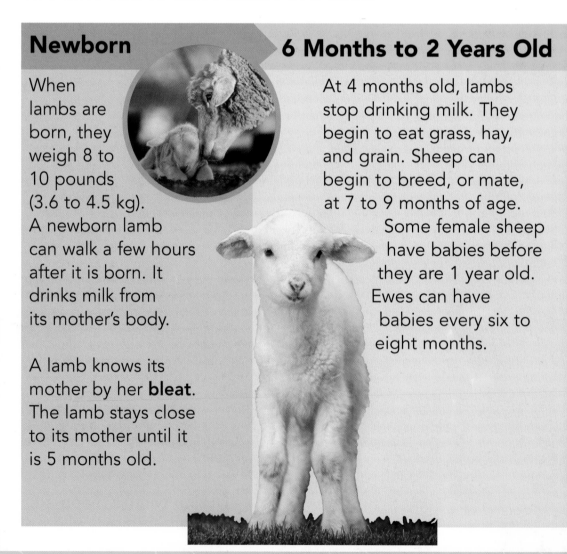

## Newborn

When lambs are born, they weigh 8 to 10 pounds (3.6 to 4.5 kg). A newborn lamb can walk a few hours after it is born. It drinks milk from its mother's body.

A lamb knows its mother by her **bleat**. The lamb stays close to its mother until it is 5 months old.

## 6 Months to 2 Years Old

At 4 months old, lambs stop drinking milk. They begin to eat grass, hay, and grain. Sheep can begin to breed, or mate, at 7 to 9 months of age. Some female sheep have babies before they are 1 year old. Ewes can have babies every six to eight months.

Most ewes give birth to twins. Lambs are often born in the spring, when the weather is warm and the grass is green.

Sheep usually live for about 10 to 12 years. With special care, some sheep can live up to 20 years.

## Adult

Sheep grow until they are 3 years old. Full-grown rams weigh between 165 and 440 pounds (75 and 200 kg). Adult ewes weigh between 110 and 330 pounds (50 and 150 kg). Sheep are between 2 and 4 feet (0.6 and 1.2 meter) tall.

# Caring for Sheep

Sheep need special care. At least once each year, sheep must be given their **vaccinations**. Sheep can become sick with many types of illnesses. Sheep often get soremouth. It causes sores on the hooves, legs, and face. It is important to check for signs of illness each day. Ill sheep should see an animal doctor, called a veterinarian.

To keep warm in winter, sheep need their wool. In the summer, the wool would keep them too warm, so sheep need to be **sheared** each spring. All of their fleece is cut off. Shearing does not hurt the sheep, but only someone who has special training should shear a sheep.

## Useful Websites

To learn more about sheep and how to care for them, visit:
**www.sheep101.info**.

If sheep are not sheared in the spring, they will be too hot in the summer and have trouble moving around.

# Myths and Legends

Sheep are found in many countries. People around the world tell stories about sheep. In the southwest United States, Navajo American Indians breed sheep. They believe that sheep are sacred animals. For more than 400 years, they have raised a special breed of sheep called the Navajo-Churro. The Navajo believe that this sheep was given to them by their gods.

Ancient Egyptians **worshiped** a god that had a sheep's head on a human body. The ancient **Celts** thought sheep symbolized war. They believed sheep helped them win battles.

Sheep have been featured in many paintings from the 5th to the 17th centuries.

# Aries the Ram

*Aries is a constellation of stars in the sky in the shape of a ram. This is a story about how the constellation was created.*

Athamas, King of Croneus, married a woman named Nephele. They had two children together. Their names were Phrixus and Helle. As a gift, Athamas gave Nephele a winged ram.

Soon Athamas took a second wife named Ino. Ino hated Nephele's children and wanted to harm them. To keep her children safe, Nephele sent them away on the back of the winged ram. The ram flew east. As the ram and the children were gliding through the air, Helle fell off. She fell right between Europe and Asia. Phrixus did not fall off and arrived safely.

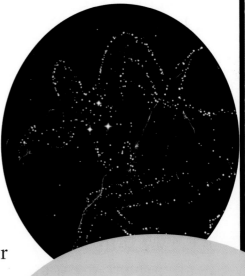

Zeus, the ruler of the Greek gods, decided to put the figure of the ram in the sky as a constellation. This was to honor the ram for its bravery, and to never forget Nephele, Helle, or Phrixus.

The Aries constellation is best viewed in the winter.

# Frequently Asked Questions

## How can I keep track of the sheep in my flock?

**Answer:** You can use tags to keep track of your sheep. A tag is usually attached to a sheep's ear. Tags are plastic or metal and have information about a sheep's age, health, parents, and farm.

## How can I keep my sheep healthy?

**Answer:** To keep sheep from becoming sick, they should be given vaccinations. Sheep should also have plenty of living space. This space must be kept clean. Sick sheep should be kept away from the rest of the flock.

## How do I get my sheep sheared?

**Answer:** For small flocks, you can bring your sheep to the shearer. You can also get together with other small flock owners and pay the shearer to come to your farm. You could learn to shear, too.

# Puzzler

See if you can answer these questions about sheep.

1. How many sheep breeds are there?
2. Where did sheep first come from?
3. What can sheep be used for?
4. What do sheep eat?
5. What are baby sheep called?

**Answers:** 1. About 800 2. Asia 3. Meat, wool, soap, oil, cheese 4. Grass, weeds, young plants, hay, corn, wheat, oats, barley, clover, twigs 5. Lambs

# Find Out More

There are many more interesting facts to learn about sheep. If you would like to learn more, take a look at these books.

Damerow, Gail. *Barnyard in Your Backyard: A Beginner's Guide to Raising Chickens, Ducks, Geese, Rabbits, Goats, Sheep, and Cows.* North Adams, MA: Storey Publishing, LLC, 2002.

Miller, Sara Swan. *Sheep.* CT: Children's Press, 2000.

# Words to Know

**adapted:** adjusted to the natural environment

**bleat:** the sound a sheep makes

**breeds:** groups of animals that have common features

**Celts:** ancient group of people who lived in western and central Europe

**climate:** weather

**gourmet:** fancy or expensive food

**grazing:** eating grass in a field

**hardy:** able to survive in a harsh environment

**mammals:** animals that have warm blood and feed milk to their young

**mouflon:** small European sheep that live in nature

**muzzles:** the nose and jaw area of animals

**predators:** animals that hunt other animals for food

**settlers:** people who go to live in a new place

**sheared:** cut or trimmed

**trade:** the act of buying and selling

**urial:** bearded reddish sheep from southern Asia

**vaccinations:** medicines that prevent disease

**worshiped:** prayed to

# Index